California State Library
Manual of Recommended Practice

Seismic Safety Standards For Library Shelving

Prepared By:

John A. Shelton
Supervising Structural Engineer
California Office of the State Architect

Approved By:
Larry D. Guthrie, Chief
Architecture and Engineering
California Office of the State Architect

California State Library Foundation
Sacramento
1990

ISBN 0-929722-39-6

PREFACE AND ACKNOWLEDGEMENT

The failure of book stacks in both public and academic libraries in the October 1987 Whittier earthquake demonstrated the need for proper design and installation of book stacks to cope with California's volatile geology.

The Uniform Building Code (UBC) has no specification for library shelving. However, the State Building Code (SBC), (Title 24, Code of California Regulations) does have a book stack specification to meet seismic standards. For many years, I have required that the SBC standard be used in public library construction projects built with federal Library Services and Construction Act (LSCA), Title II funds. This manual will assist those LSCA projects in conforming to the SBC specification.

Of course, the manual is not limited to LSCA, Title II. The SBC standard will also be required for projects to be built with funds from the California Library Construction and Renovation Bond Act of 1988 (Proposition 85), and this manual will also assist those projects. I also encourage every California Architect and Librarian involved in planning and installing library shelving, whether it be school, hospital, college or public libraries, and whether the project be in the private or public sector, to adopt the practices in this manual.

Credit for the idea of this manual goes to Nancy McAdams, of McAdams Planning Associates, Austin, Texas. Ms. McAdam's work as a library consultant to many California research library projects, including the State Library Annex project, led her to suggest the idea to Cy Silver, State Library Facilities Planning Consultant. Mr. Silver approached Michael Bocchicchio, Sr., A.I.A., State Architect who assigned John Shelton, Supervising Structural Engineer, to the project. It is Mr. Shelton who did the work, and to whom thanks are due.

The library shelving designed and installed in conformance with the practice recommended in this manual is expected to withstand the range of earthquakes commonly encountered in California, and by so doing both protect the people who work in and use libraries from injury from collapsing book stacks, and protect the local budgets from the cost of replacing such damaged or destroyed book stacks.

Gary E. Strong
California State Librarian
September 1989

CONTENTS

MANUAL OF RECOMMENDED PRACTICE

SEISMIC SAFETY STANDARDS FOR LIBRARY SHELVING

1. INTRODUCTION:

1.1 This Manual of Recommended Practice describes the practices to be followed for installation of, or modifications to, library shelving. There is no absolute guarantee for the complete safety of personnel or library property during and after an earthquake because of the unpredictable magnitude and nature of the forces generated during a significant seismic event. However, adherence to the practices stated herein will minimize the potential for injury of people from library shelving during an earthquake and maximize the protection of property.

The term "Seismic Safety" as used in this manual relates to the ability of the installed shelving and their supports to resist overturning or collapse and, therefore, prevent injury to people and damage to property during a seismic event.

1.2 This manual is limited to the support structures and bracing for library shelving. This applies to either new or existing shelving in either new or existing facilities.

1.3 The development of this manual has been based on the physical characteristics of existing shelving.

2. STATEMENT OF POLICY:

2.1 Since the State Library requires that library shelving installed in projects using funds from the federal Library Services and Construction Act or the state California Library Construction and Renovation Bond Act of 1988 (Proposition 85) conform to the standards in Part 2 of the SBC, (Title 24, Code of California Regulations), it is the policy of the State Library that all such shelving be designed and installed in conformance with this manual.

The State Library also encourages all jurisdictions and agencies responsible for libraries to adopt the requirements of this manual. Those libraries would include not only public libraries, but also school, special and academic libraries, in both the private and the public sector.

2.2 This policy applies to all new facilities and to the retrofit or upgrading of existing facilities.

2.3 No exceptions to this policy may be granted. However, when special facility site conditions prevent conformance to specific details or instructions contained in this manual the State Library, with the concurrence of the Office of the State Architect, may grant waivers on an individual project basis.

3. DESIGN CONSIDERATIONS:

3.1 Assumptions:

The primary assumption that has been made in the formulation of this manual is that the buildings or other structures which house the shelving will themselves be able to withstand the horizontal and vertical loads generated during an earthquake. This assumption applies to both new and existing buildings or structures. The responsible Architect or Engineer shall assure the adequacy of all building components which will support library shelving to withstand the seismic loads as defined in the 1988 California Building Standards Code - Title 24 CCR.

3.2 Alternatives Considered:

Several alternatives were examined during the study conducted in the preparation of this manual. Briefly, these alternatives were focused on:

a. Location and magnitude of loads (shelving completely or partially filled in various combinations).

b. Structural arrangements for independently stable shelving (steel, wood or concrete bases; grouped shelving, etc.).

c. Structural arrangements for shelving partially or totally supported by building (floor bolting, framing to walls, support from ceiling, etc.).

The results of the alternative studies are in this manual's Appendixes A and B.

3.3 Design Criteria:

The design criteria are stated in detail in Appendix A to this manual. These criteria presents the design requirements which should be followed for library facilities.

Seismic loads may be evaluated for library shelving supports through calculation of the equivalent static forces. Formulas to determine these equivalent static forces are detailed in Appendix A to this manual for both rigidly supported and flexibly supported racks or equipment.

The support system and connections shown on the suggested detail drawings have been designed to meet all applicable requirements of Part 2, Title 24, CCR, for Seismic Zones 3 and 4, to resist the seismic forces generated by the masses of the library materials and the support system. Appendix A also lists restrictions to the design which must be adhered to.

4. DESIGN OF NEW FACILITIES:

4.1 All new library facilities should be designed in conformance with the design criteria, standard details and installation instructions contained in Appendixes A and B of this manual. This data applies to the installation and support of library shelves. The data may also be adapted for use with other shelving not specifically identified.

4.2 The recommended practices for design of library shelves are based on different sizes of shelves. The shelving may vary in height, width and length dependent on the materials stored and size of rooms used.

5. UPGRADING OF EXISTING FACILITIES:

5.1 All existing library shelving should be upgraded to conformance with the design criteria, standard details, and installation instructions contained in Appendixes A and B of this manual. This data applies to all new shelving and/or the revisions necessary to brace or support existing shelving. Braced-frame stacks shall be inspected to determine the adequacy of the braces and their connections. An architect or engineer should be consulted for the best method appropriate for any installation.

5.2 In some cases it will be necessary to modify the standard details to fit non-standard size shelves. Such modifications shall be made on a case study basis, shall conform to the design criteria in Appendix A, and shall conform to the intent of the standard details and installation instructions.

APPENDIX "A"

DESIGN CRITERIA

1. PURPOSE AND SCOPE

These criteria prescribe the standards for design of anchorages, supports, and installation of library shelves throughout California. The intention of these requirements is that the installed shelving will remain in place during and after a seismic disturbance.

Seismic design of shelving supports and details to stabilize them will be subject to the constraints and special requirements contained in the 1988 California Building Standards Code - Title 24 CCR.

Seismic forces used for design of shelving supports are comparatively higher than similar forces used in the design of buildings mainly because the shelf supports lack the extra margin of safety provided by reserve strength mechanisms found in buildings, such as participation of architectural elements, inelastic behavior of architectural elements, and redundancy in structural systems; also, shelves located high in buildings are subjected to amplified ground forces as a result of the dynamic response of the building.

Floor supported library shelving less than five feet in height (including contents) does not require anchorage, if the width of the supporting base is greater than two-thirds the height.

2. SEISMIC FORCES

2.1 General

In general, horizontal and vertical components of seismic forces are applied simultaneously to determine the stability and strength requirements of a structure. The vertical component of the seismic loads involved in design of the shelf supports shall be applied as stipulated in the 1988 California Building Standards Code - Title 24 CCR.

The horizontal forces are a function, among other variables, of the type of restraints operating on the shelf supports; these restraints can offer rigid or flexible supports.

For the purpose of support and anchorage design of shelves, the following definitions will apply:

a. Rigidly Supported Shelves

Shelf supports which are laterally braced at the top and anchored to the floor shall be considered rigidly supported. Wood shelves with full width 3/4" plywood end panels anchored to the floor shall be considered rigidly supported.

b. Flexibly Supported Shelves

Shelf supports anchored to the floor as cantilevers and not laterally braced will be considered flexibly supported unless it can be shown that the fixed base period is less than .06 seconds.

2.2 Design Seismic Forces

The equivalent static forces to be used for design shall be assumed to act in any horizontal direction and applied at the center of gravity of the loads. They will be calculated by the following formulas:

a. <u>Rigidly Supported on Ground Floor or Below</u>

F_p = (2/3) Z I Cp Wp

b. <u>Flexibly Supported on Ground Floor or Below</u>

F_p = Z I Cp Wp

c. <u>Rigidly Supported in Higher Floors</u>

F_p = Z I Cp Wp

d. <u>Flexibly Supported in Higher Floors</u>

F_p = (4) Z I Cp Wp for support systems with period > .06 seconds.

Fp = (2) Z I Cp Wp for support systems with period < .06 seconds.

Where:

F_p = Equivalent static design seismic force
Z = .30 for Zone 3, .40 for Zone 4. Each site shall be assigned to a seismic zone in accordance with Figure No. 2, Chapter 23, of the 1988 California Building Standards Code - Title 24 CCR.
I = 1.5 for Essential Services and Hospital buildings
I = 1.0 for all other buildings
Cp = 0.75
Wp = Total Weight of Shelving and contents. The minimum vertical design live load shall be 50 psf per shelf or 50 pcf per total volume of the shelves whichever is greater.

The component anchorage shall be designed for the horizontal force, Fp, acting simultaneously with a vertical seismic force equal to one-third the horizontal force, Fp.

Floor-supported book stacks not more the 5 feet in height need not be anchored if the width of the supporting base or width between the exterior legs is equal to or greater than two-thirds the height.

3. <u>SUPPORT SYSTEM AND CONNECTIONS</u>

The support system and connections shall be designed and constructed to meet all applicable requirements of 1988 California Building Standards Code - Title 24 CCR, to resist the seismic forces generated by the masses of the shelf contents and their support system.

In addition, the following restriction shall be adhered to:

a. Allowable Working Stress

One-third increase in allowable working stress is permitted for member design. One-third increase is not permitted for welded, bolted, or other connections.

b. Securing of Shelving and Shelving Supports

Use of friction as a method of resisting seismic forces will not be allowed. Shelving and shelving supports shall be secured by bolts, screws, or other acceptable positive means of resisting seismic forces.

c. Anchor Bolts

Expansion type anchors shall not be used for connections in direct tension to concrete and masonry for other than seismic loads. Tension connections shall be made with anchor bolts embedded, epoxy grouted at least eight (8) bolt diameters into the concrete, or masonry anchors of a bearing type undercut anchor installed in a hole with a conical undercut near its blind end. Anchor bolts embedded in grouted masonry shall have at least 1 inch of grout between the bolt and the masonry. Expansion type anchors may be used for anchorage to concrete or masonry when the tension loads are required to resist seismic loads only. The depth of embedment of expansion anchors in masonry shall not include the dimension of the face shell.

d. Sizes of Screws and Bolts

Screws and bolts for connections shall meet the minimum size requirements as follows:

Wood connections	- 14 gauge wood screws 1/4 inch diameter bolts
Metal connections	- 1/4 inch diameter bolts
Concrete connections	- 1/4 inch diameter bolts
Masonry connections	- 1/4 inch diameter bolts

e. Unbraced Length and Radius of Gyration of Bracing Members

Maximum allowable L/r values shall be as follows:

300 for tension members other than rods
200 for compression members

APPENDIX "B"

STANDARD DETAILS
AND
INSTALLATION INSTRUCTIONS

GENERAL NOTES

MATERIAL:

1. ALL STEEL ELEMENTS SHALL BE ASTM A36 OR BETTER.

2. LUMBER SHALL BE DOUGLAS FIR-LARCH, DENSE #2 OR BETTER.

3. PLYWOOD SHALL BE B-B EXT-APA WITH EXTERIOR GLUE (CDX) OR BETTER.

4. FABRICATION AND INSTALLATION OF STEEL SHALL CONFORM TO THE APPLICABLE SECTIONS OF AISC SPECIFICATIONS AND 1988 CALIFORNIA BUILDING STANDARDS CODE -TITLE 24, CCR.

5. BOLTS SHALL BE ASTM A307 MACHINE BOLTS WITH HEX HEAD AND NUT. MINIMUM BOLT DIAMETER SHALL BE 1/4 INCH.

6. LAG SCREWS SHALL BE STEEL AND SHALL BE SCREWED INTO PRE-DRILLED HOLES WHICH ARE 75 PERCENT OF THE DIAMETER OF THE ROOT OF THREAD, ENLARGED TO SHANK DIAMETER FOR THE LENGTH OF THE SHANK.

7. WOOD SCREWS SHALL BE STEEL AND SHALL BE SCREWED INTO PRE-DRILLED HOLES WHICH ARE 70 PERCENT OF THE DIAMETER OF THE ROOT OF THE THREAD.

8. LUMBER AND PLYWOOD SHALL BE CUT AS REQUIRED TO FIT THE INSTALLATION AND PAINTED TO MATCH THE FINISHES OF ADJACENT EQUIPMENT AND/OR SUPPORTS.

9. ANCHOR BOLTS IN CONCRETE OR MASONRY WALLS SHALL BE GROUTED AT LEAST 8-BOLT DIAMETERS INTO WALL, CAST-IN-PLACE, OR EXPANSION ANCHORS. ALL ANCHOR BOLTS IN MASONRY SHALL HAVE AT LEAST 1 INCH OF GROUT BETWEEN THE BOLT AND THE MASONRY.

10. EXPANSION ANCHORS SHALL BE "HILTI" KWIK-BOLTS OR APPROVED EQUAL. ALL EXPANSION ANCHORS SHALL HAVE ICBO APPROVAL.

11. EXPANSION ANCHORS IN MASONRY SHALL BE "PHILLIPS RED HEAD" OR APPROVED EQUAL.

12. WELDING ELECTRODES SHALL BE E70XX.

Office of the State Architect

INSTALLATION:

SHELF BASE ANCHORS:

LIBRARY SHELVES SHALL BE SUPPLIED WITH THE REQUIRED ANCHOR BOLTS FOR THE BASE CONNECTION. EVERY ANCHOR BOLT SUPPLIED SHALL BE INSTALLED PER APPROVED ENGINEERING DRAWINGS.

IN GENERAL, FOR INSTALLATION ONTO CONCRETE FLOOR SLABS, EXPANSION TYPE ANCHORS OF THE LARGEST DIAMETER THAT FITS INTO THE PRE-PUNCHED HOLES IN THE BASE ANGLES SHALL BE USED. VERIFY LOCATION OF STEEL IN PRESTRESSED CONCRETE BEFORE DRILLING. FOR INSTALLATION ONTO WOOD FLOORS, LAG SCREWS OF EQUAL DIAMETER SHALL BE USED. THESE LAG SCREWS SHALL BE CONNECTED TO THE FLOOR JOISTS OR BLOCKING.

DEVIATION:

THIS SET OF DOCUMENTS AND DRAWINGS PROVIDE THE BRACING SYSTEM AND CONNECTION REQUIREMENT FOR A TYPICAL INSTALLATION.

ANY DEVIATION REQUIRED TO FIT AN UNUSUAL SITE CONDITION SHALL BE DESIGNED AND SUBMITTED FOR APPROVAL. FABRICATION AND INSTALLATION MAY NOT BEGIN UNTIL APPROVAL HAS BEEN OBTAINED.

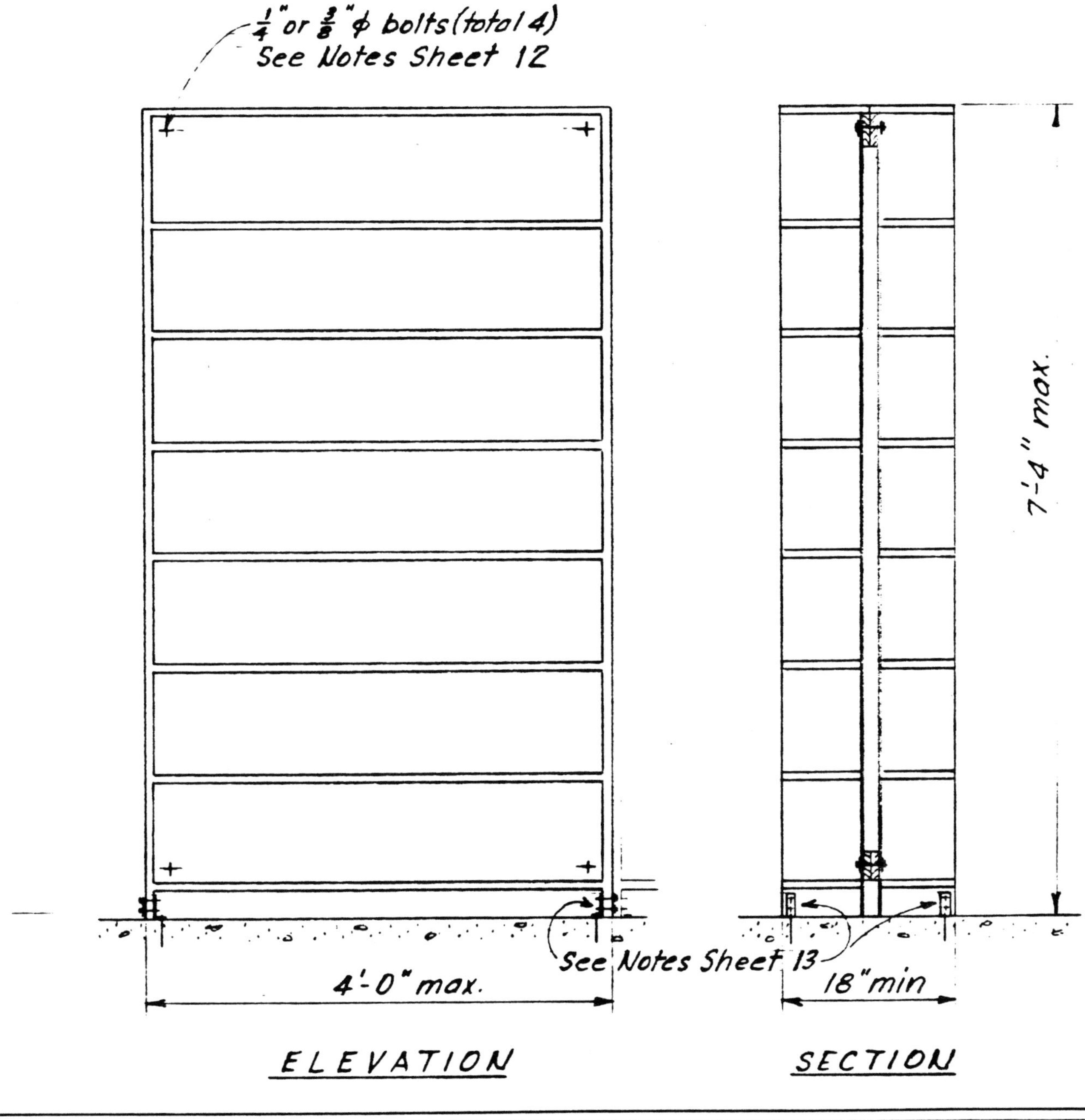

Office of the State Architect

MICHAEL J. BOCCHICCHIO SR., A.I.A. · STATE ARCHITECT

WOOD SHELVES
FREE STANDING
BASE ANCHORED

Drawn by JAS

12

88 48927

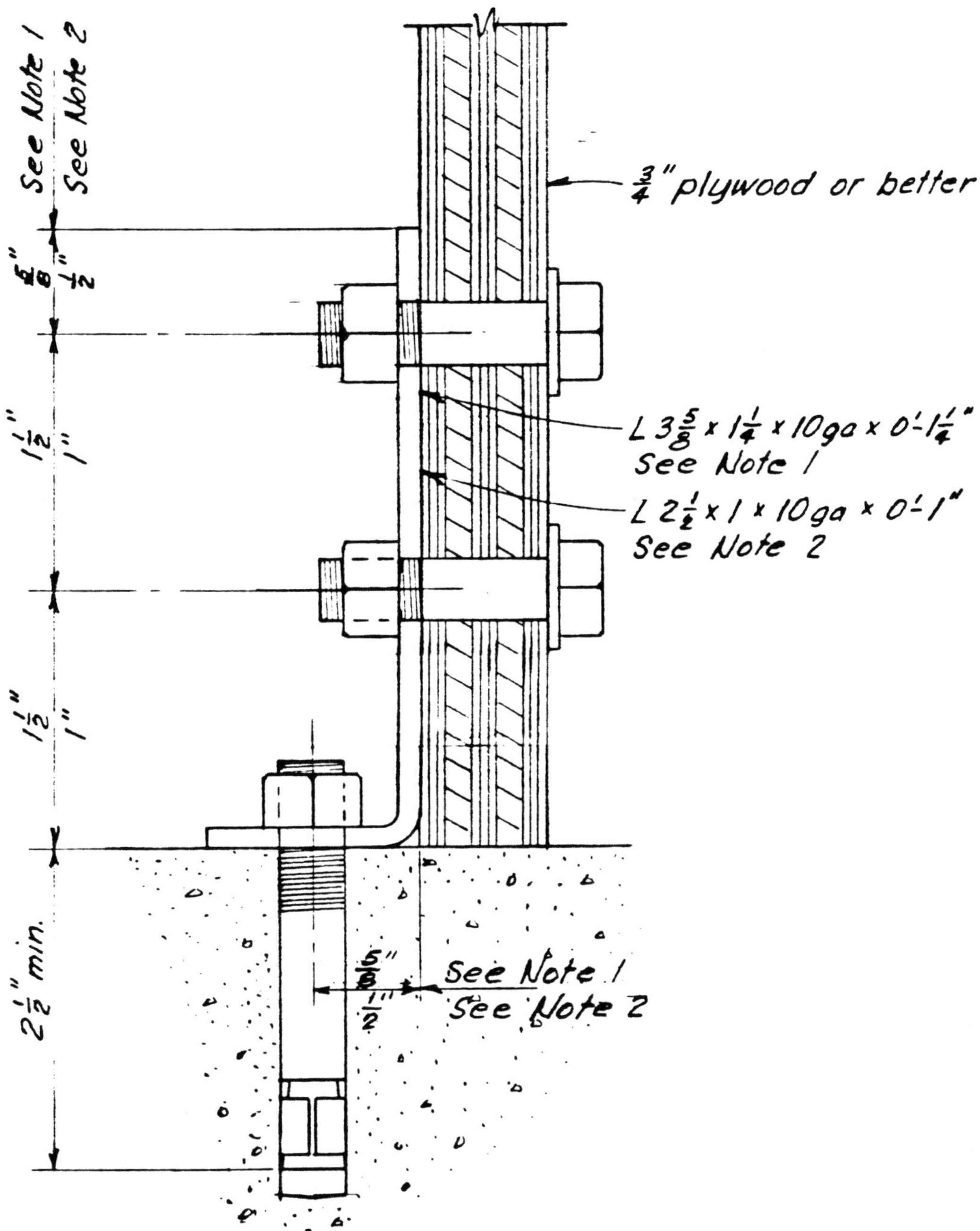

NOTES:

1. Above ground floor - 3/8" ϕ bolts or anchors.
2. At ground floor or below - 1/4" ϕ bolts or anchors

Office of the State Architect MICHAEL J. BOCCHICCHIO SR., A.I.A. · STATE ARCHITECT	WOOD SHELVES BASE ANCHOR	Drawn by JAS	13

86 48927

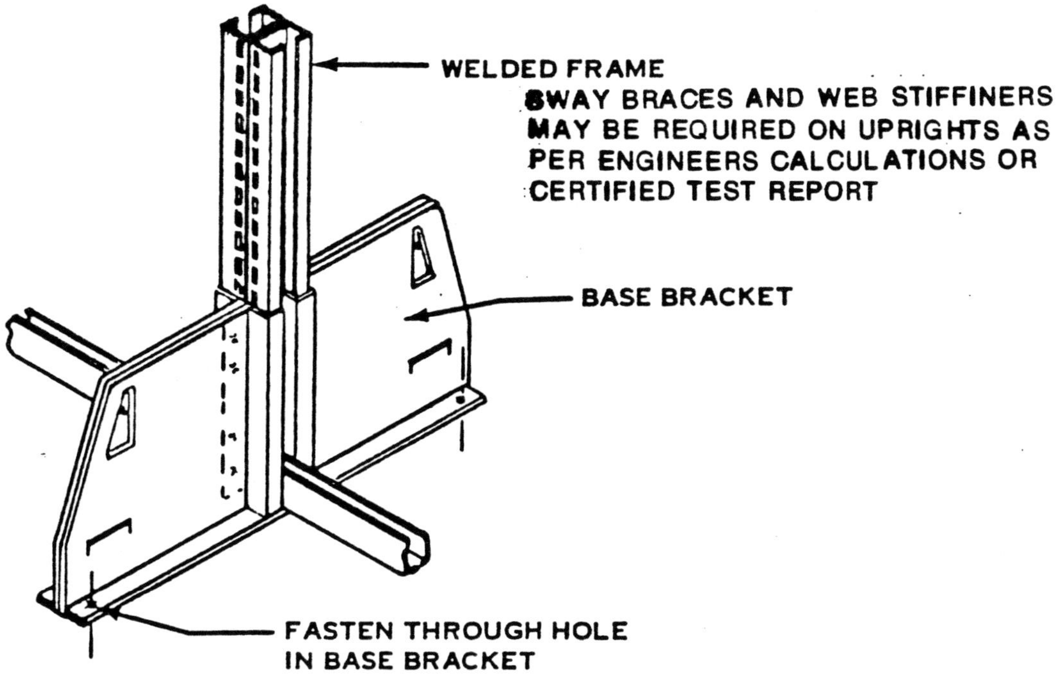

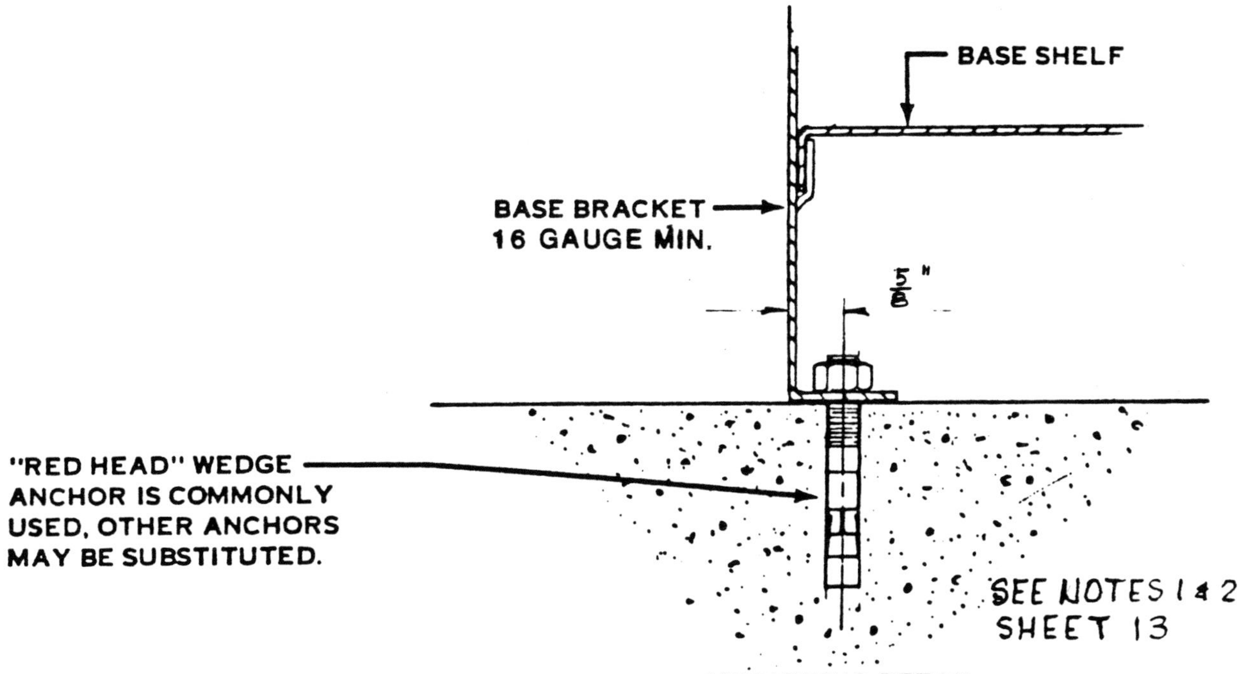

ANCHORING DETAIL

Office of the State Architect MICHAEL J. BOCCHICCHIO SR., A.I.A. · STATE ARCHITECT	STEEL SHELVES BASE ANCHOR	Drawn by JAS	14

88 48927

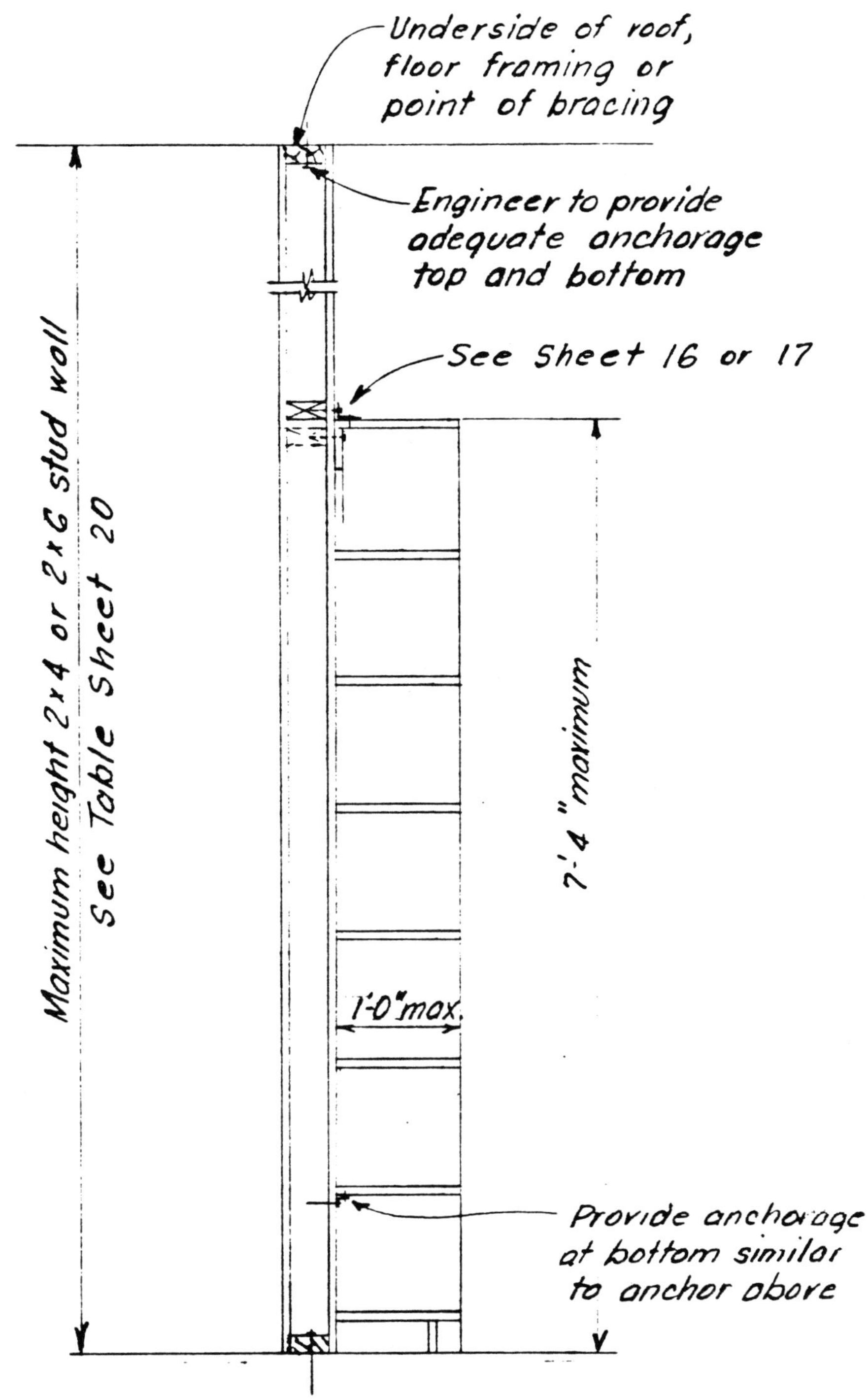

Office of the State Architect
MICHAEL J. BOCCHICCHIO SR., A.I.A. · STATE ARCHITECT

LIBRARY SHELVES ANCHORED TO WOOD STUD WALL

Drawn by JAS

15

88 48927

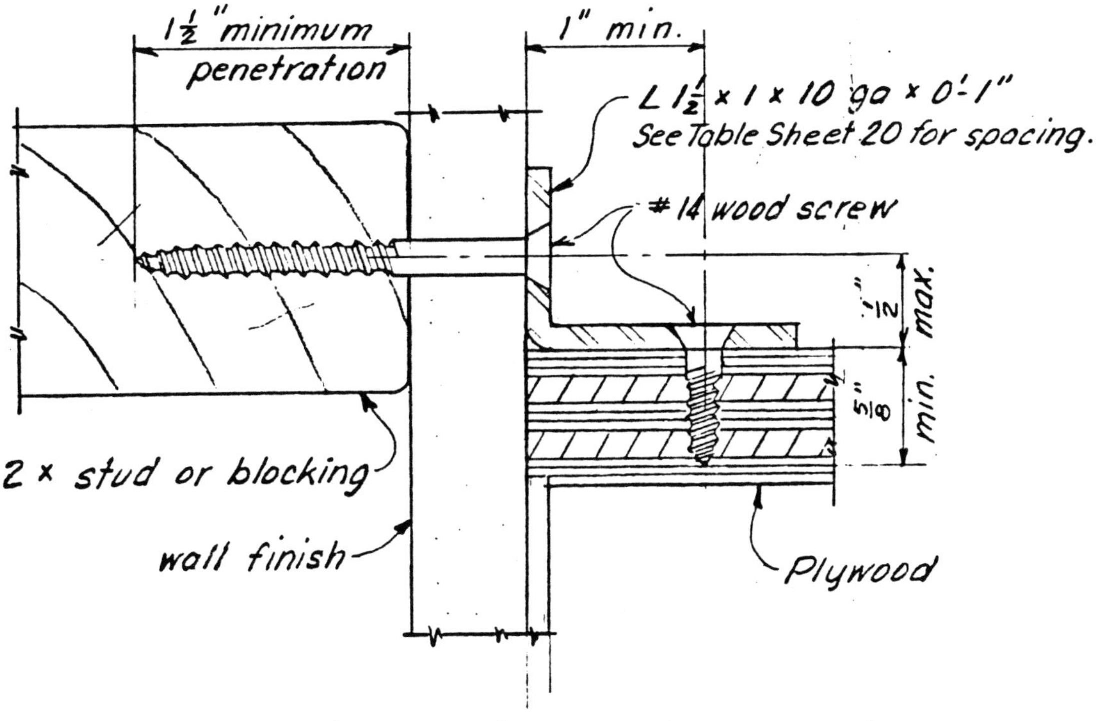

Note: For Steel Shelves use ¼" ϕ bolt in lieu of #14 wood screw

Office of the State Architect
MICHAEL J. BOCCHICCHIO SR., A.I.A. · STATE ARCHITECT

WOOD SHELVES
TOP ANCHOR
WOOD STUDS

Drawn by JAS

16

BB 48927

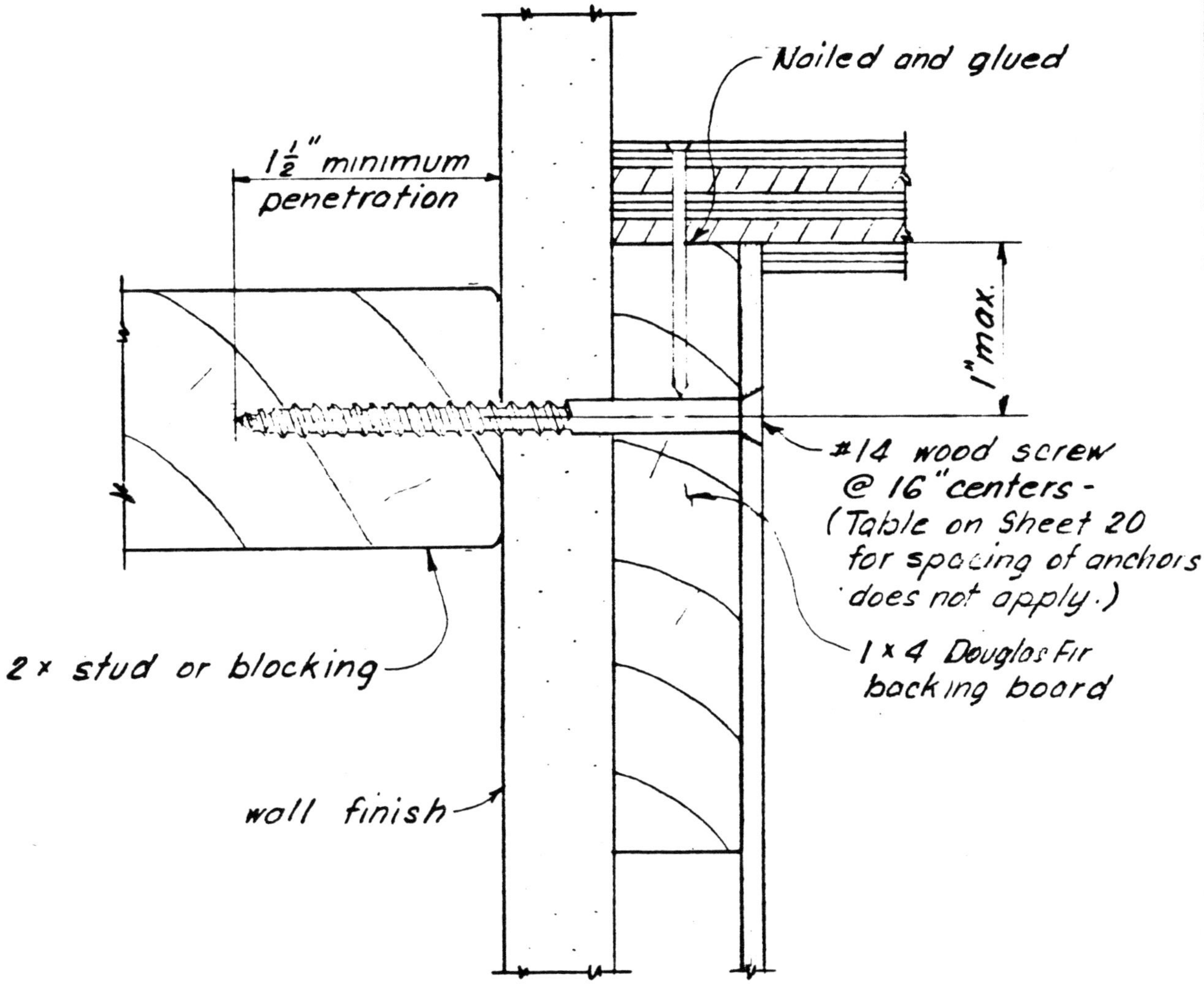

Office of the State Architect
MICHAEL J. BOCCHICCHIO SR., A.I.A. · STATE ARCHITECT

WOOD SHELVES
TOP ANCHOR
WOOD STUDS

Drawn by JAS

17

88 48927

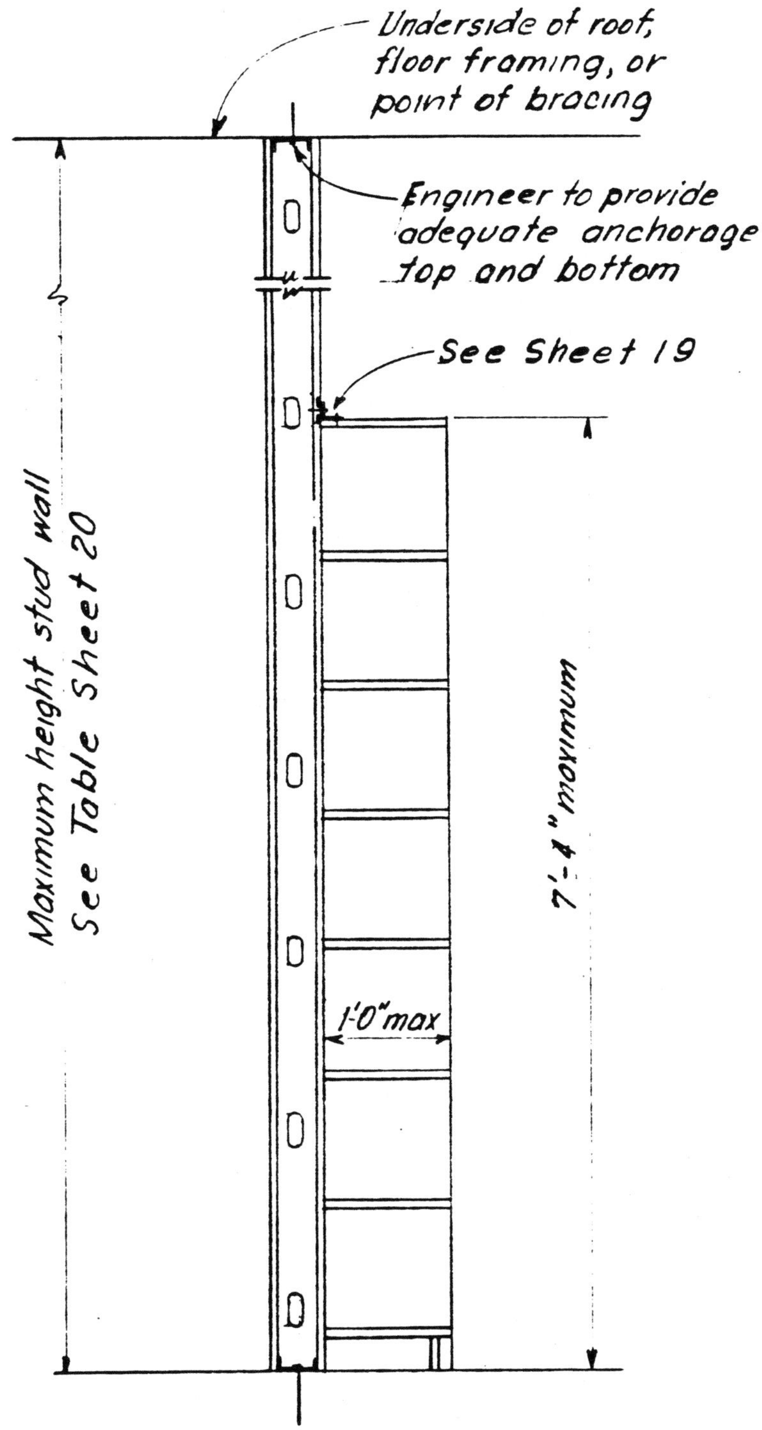

Office of the State Architect
MICHAEL J. BOCCHICCHIO SR., A.I.A. · STATE ARCHITECT

LIBRARY SHELVES ANCHORED TO STEEL STUD WALL

Drawn by JAS

88 48927

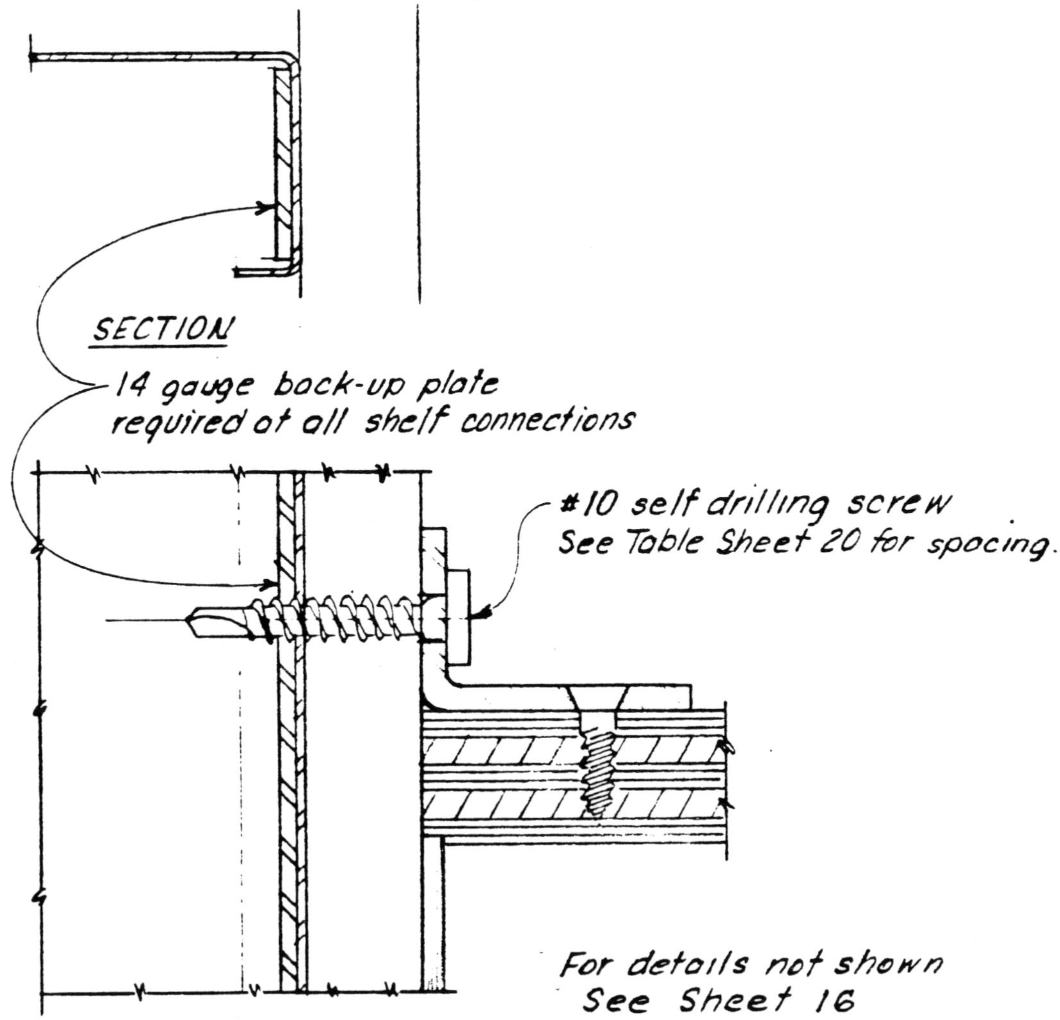

Office of the State Architect
MICHAEL J. BOCCHICCHIO SR., A.I.A. · STATE ARCHITECT

WOOD SHELVES
TOP ANCHOR
STEEL STUDS

Drawn by JAS

88 48927

MAXIMUM HEIGHT NON BEARING WALLS (IN FEET)

WOOD STUDS (CONSTRUCTION GRADE)

DISTANCE BETWEEN ANCHORS	16		24		32		48	
SEISMIC ZONE	3	4	3	4	3	4	3	4
2 X 4 @ 16	14	13	12	11	11	9.5	9	8
2 X 6 @ 16	20	20	20	20	20	18	17	14

STEEL 3-5/8" X 1-1/4" DRYWALL C-STUD @ 16" (.25-.42 LIP)

DISTANCE BETWEEN ANCHORS	16		24		32		48	
SEISMIC ZONE	3	4	3	4	3	4	3	4
18 GAUGE	18	16	15	14	13	12	11	10
20 GAUGE	16	14	13	11	11	10	10	9
22 GAUGE	14	12	11	10	10	9	8	8
25 GAUGE	10	9	9	8	8	*	*	*

STEEL 3-5/8" X 1-3/8" WIDE-FLANGE CHANNEL STUD @ 16" (NO LIP)

DISTANCE BETWEEN ANCHORS	16		24		32		48	
SEISMIC ZONE	3	4	3	4	3	4	3	4
18 GAUGE	14	13	12	11	11	10	9	8
20 GAUGE	11	10	10	9	9	8	8	*
22 GAUGE	9	8	8	*	*	*	*	*

* Cannot be used.

Listed heights are distances between points of lateral support placed perpendicular to the plane of the wall. Increases in unsupported height are permitted where justified by an analysis.

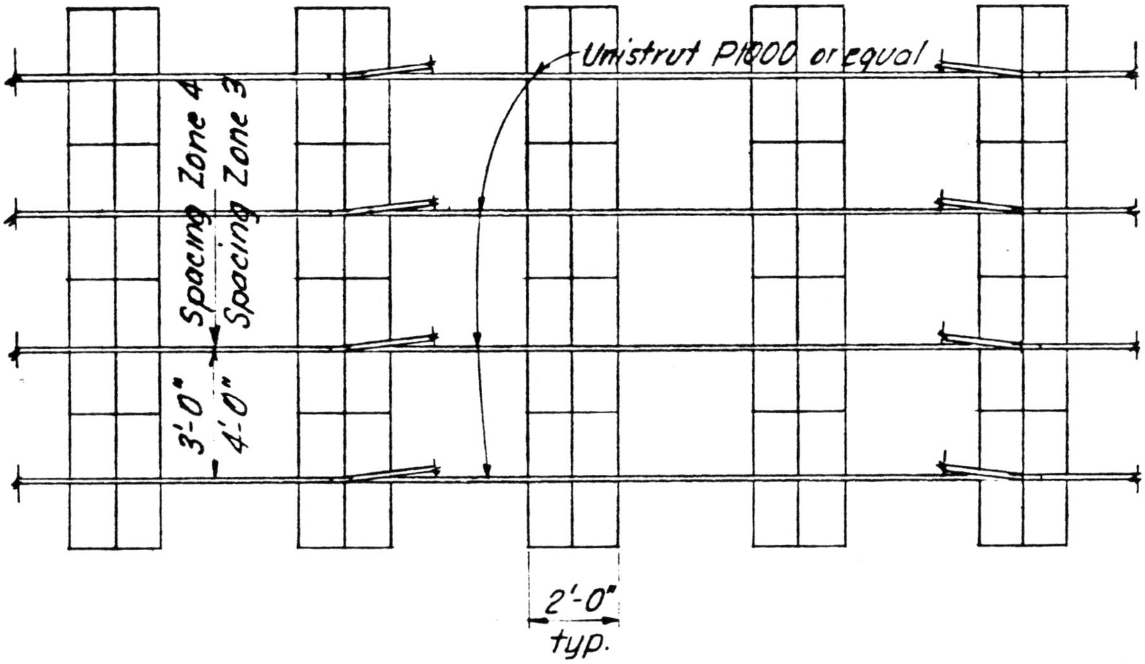

PLAN

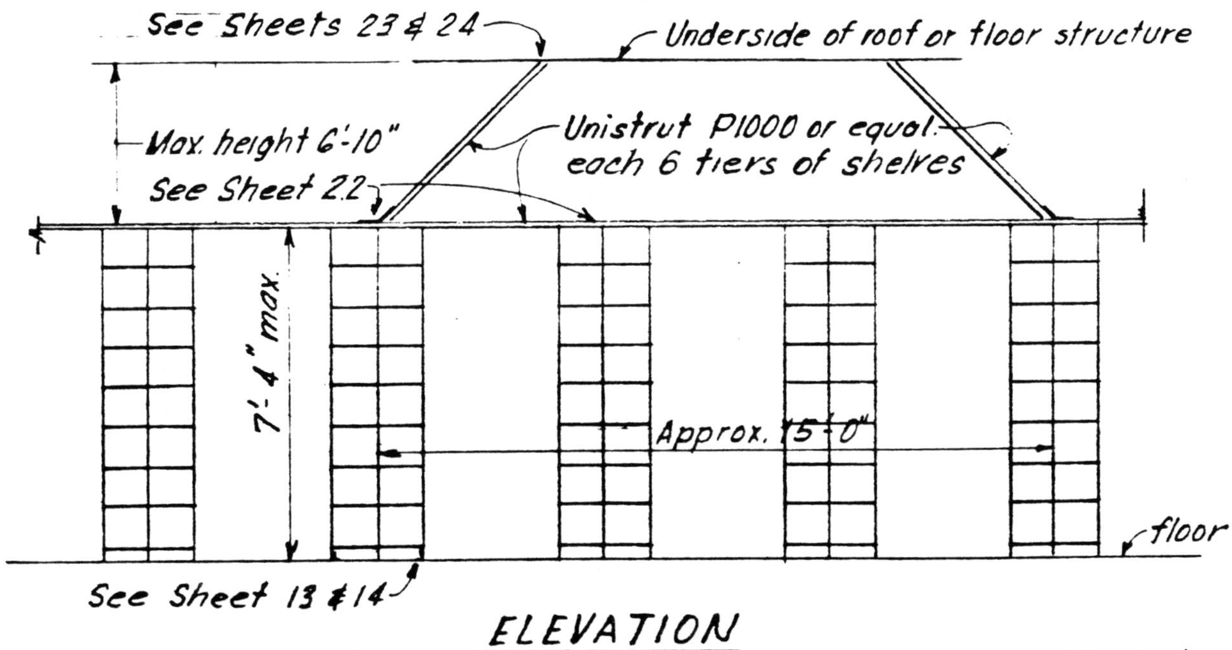

ELEVATION

Scale: $\frac{1}{8}$" = 1'-0"

Office of the State Architect MICHAEL J. BOCCHICCHIO SR., A.I.A. · STATE ARCHITECT	FREE STANDING SHELVES TOP BRACED	Drawn by JAS	21

88 48927

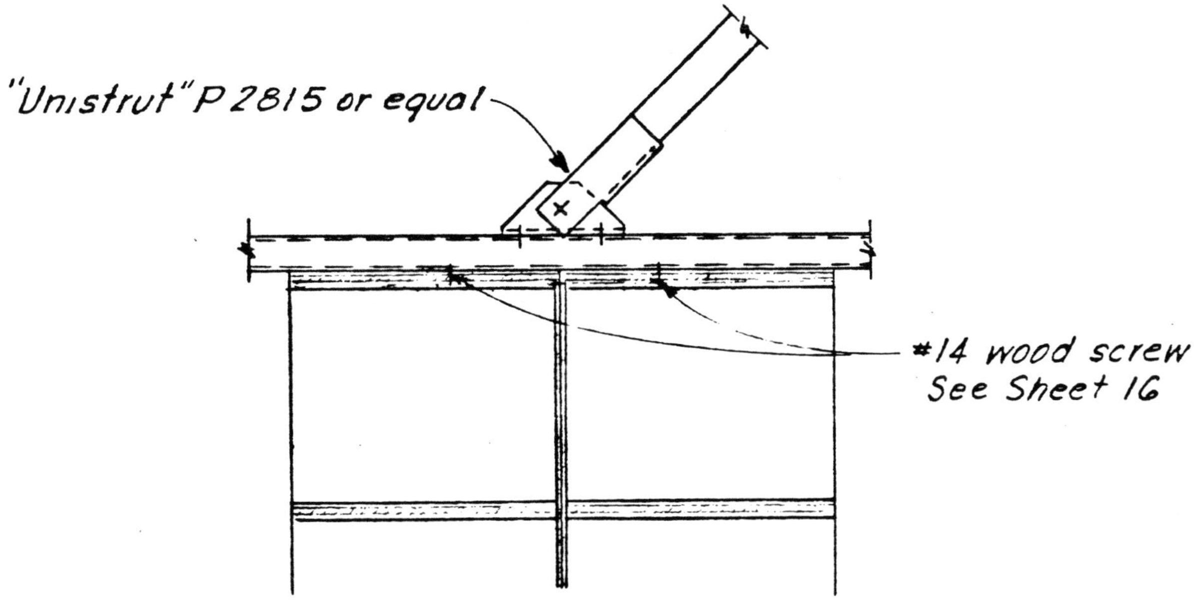

TOP OF SHELVES

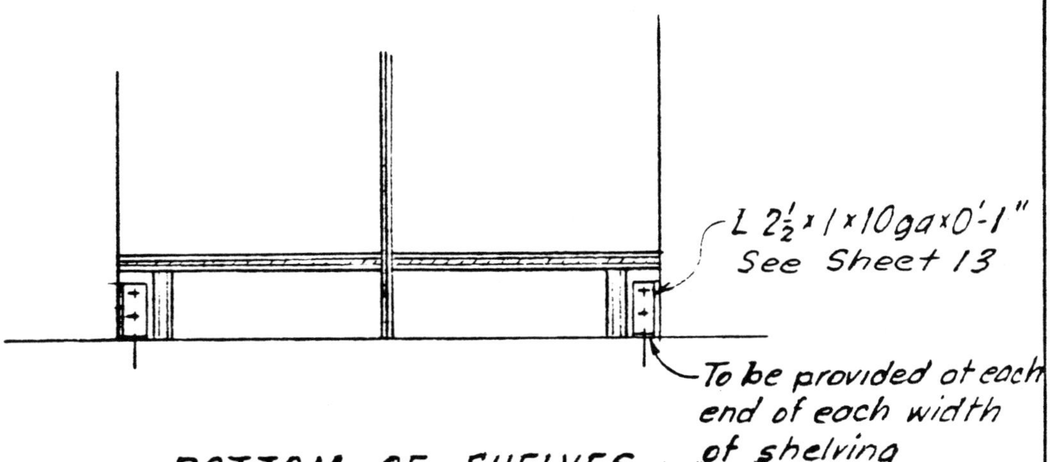

BOTTOM OF SHELVES

Office of the State Architect MICHAEL J. BOCCHICCHIO SR., A.I.A. · STATE ARCHITECT	FREE STANDING SHELVES TOP & BOTTOM CONNECTIONS	Drawn by JAS	22

88 48927

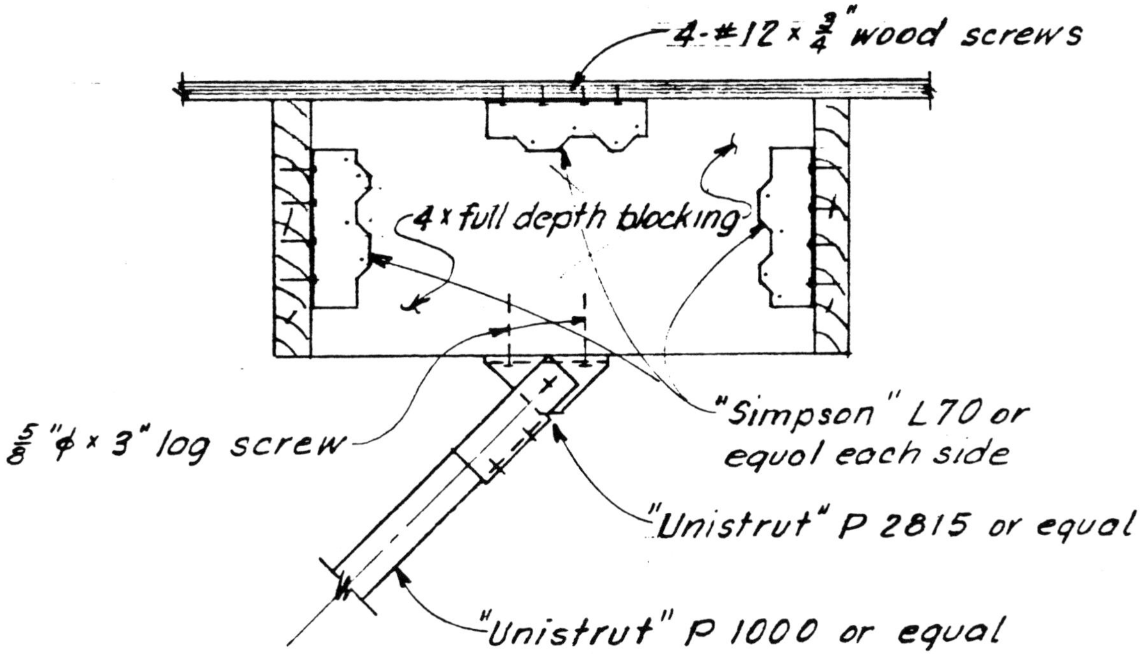

Scale: $1\frac{1}{2}$" = 1'-0"

Office of the State Architect — MICHAEL J. BOCCHICCHIO SR., A.I.A. · STATE ARCHITECT	FREE STANDING SHELVES TOP BRACE CONNECTION	Drawn by JAS	23

88 48927

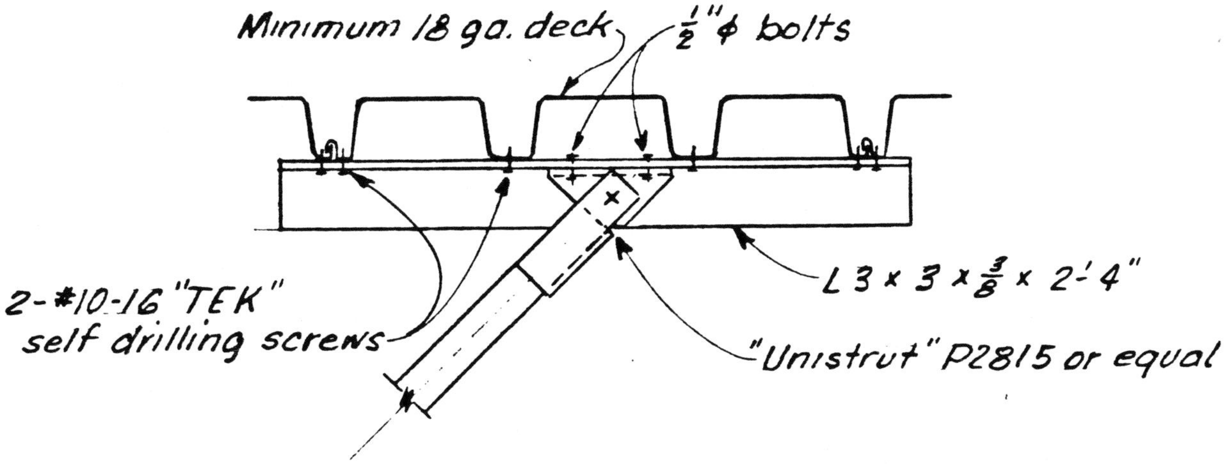

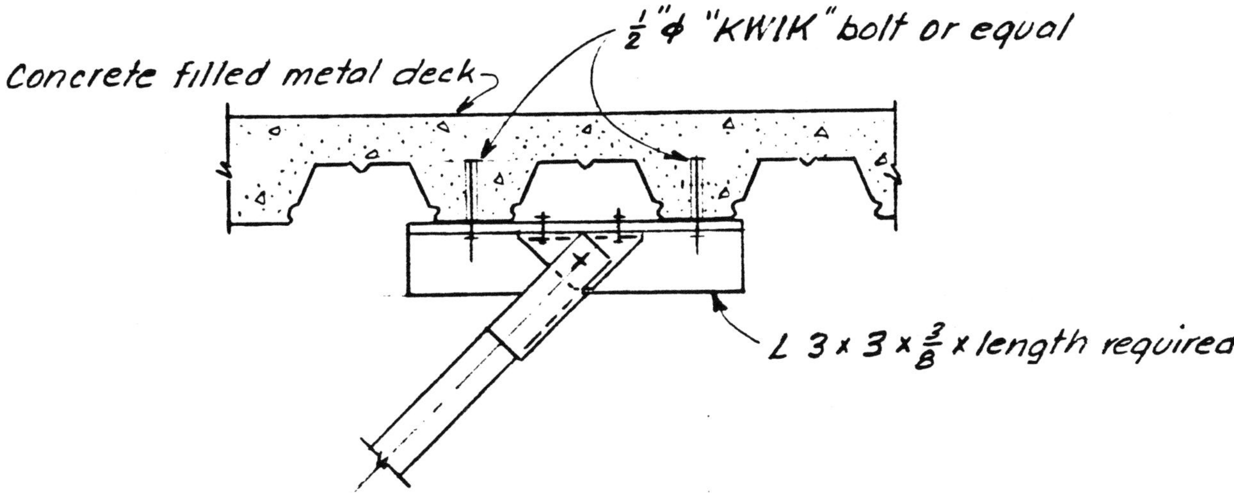

Office of the State Architect MICHAEL J. BOCCHICCHIO SR., A.I.A. · STATE ARCHITECT	FREE STANDING SHELVES TOP BRACE CONNECTION	Drawn by JAS	24

88 48927

APPENDIX "C"

SPECIFICATIONS

SPECIFICATIONS:

WOOD LIBRARY SHELVES

PART 1 - GENERAL

1.01 RELATED DOCUMENTS:

Drawings and general provisions of Contract, including General and Supplementary Conditions and Division-1 Specification sections, apply to work of this section.

1.02 RELATED WORK:

List those trades which supply backing and bracing materials for anchorage of shelving.

1.03 DESCRIPTION OF WORK:

Extent of library shelving is indicated on drawings.

Types of library shelving required includes:

Adjustable shelves.

1.04 QUALITY ASSURANCE:

Manufacturers: Provide each type of library shelving by one manufacturer for entire project.

Standards: Wood Casework, "Deluxe Grade." Use WIC "Manual of Millwork" and supplements thereto, 1988 edition, on minimum materials and construction standards.

Regulatory Requirements: Provide book stack design complying with structural and anchoring requirements for Seismic Zone (3 or 4).

1.05 ACCEPTABLE MANUFACTURERS:

Available Manufacturers: Subject to compliance with requirements, manufacturers offering products which may be incorporated in the work include, but are not limited to, the following:

RETAIN ABOVE OR BELOW, FOR NONPROPRIETARY OR SEMIPROPRIETARY SPECIFICATION.

Manufacturer: Subject to compliance with requirements, provide products of one of the following:

Library Berean, Inc.
Brodart Company
Borroughs Manufacturing Corp.
Texwood Furniture Corp.

DELETE OR ADD MANUFACTURERS TO ABOVE LIST AS DESIRED

1.06 DELIVERY:

A. Do not deliver until wet operations in building are complete and storage area is closed in and broom clean, with relative humidity 65% or less at 60F to 90F, and Equilibrium Moisture Content (EMC) conditions between 8% and 12%.

PART 2 - PRODUCTS

2.01 DEFINITIONS:

A. Exposed, semi-exposed and concealed portions of casework shall be as defined in Section 14, 1988 edition, WIC "Manual of Millwork" (WICMM).

2.02 EXPOSED MATERIALS:

A. Solid Lumber: Clear, dry, sound, selected for compatible grain and color, no defects. Species - Northern Hard Maple, White or Yellow Birch, or Red or White Oak; referenced "maple", "birch", "oak". Maple and birch may be combined in same cabinet if compatible grain and color.

B. Plywood Face Veneer: Same species as exposed solid lumber, clear, selected for grain and color compatible with exposed solid lumber, no defects. Crossbandings, solid, no voids. Edge band exposed edges with solid wood, same species as face or plastic edge molding in harmonizing color.

C. Compatible Grain and Color: Do not use exposed faces or lighter than average color joined with exposed faces of darker than average color, or two adjacent faces noticeable dissimilar in grain, figure, and natural character markings.

2.03 SEMI-EXPOSED MATERIALS:

A. Solid Lumber: Dry, sound, selected to eliminate appearance defects. Any species of hardwood, or softwood of similar color and grain to exposed portions.

B. Plywood: Hardwood ANSI/HPMA-83, "B" Grade or softwood PSA-83, "A" Grade. Species, similar color and grain to exposed portions. Edge band plywood edges.

2.04 CONCEALED MATERIALS:

A. Solid Lumber or Plywood: Any species with no defects affecting strength or utility.

B. Particleboard: Use only as core material except as specified otherwise. Comply with ANSI 208.1-89, Table 1, Grade 1-M-3. (Not to be used as shelving.)

Surfaces shall have smooth finish.

Screws in particleboard, sheet metal gimlet point type only.

2.05 WOOD FINISH:

A. General: Complete factory finish work. Minor differences between best standard finish of manufacturers named in Article 1.05 and that meeting the following requirements will not cause rejection.

B. Preparation: Machine, hand sand, and hand scrape exposed and semi-exposed portions to obtain smoothness before applying finishing materials. Machine marks, cross sanding, tool marks or similar blemishes not acceptable. Mill edges straight and true, corners broken slightly. Concealed portions have no surface smoothness requirements.

C. Exposed Portions: Finish as follows: WIC System #3 or #4 from Section 25 or 1) sealer coat if needed, sandpaper surfaces; 2) stain, as required to match color selected; 3) mineral filler for open grained wood, wipe across grain and sand; 4) sealer coat, sand and clean; 5) multiple coats first grade durable cabinet finish lacquer or synthetic resin varnish. Heat dry each coat. Sand each coat to produce smooth finished surface of even satin luster, free from imperfections.

D. Semi-Exposed Portions: Finish as follows: 1) sealer coat colored to match, sandpaper surfaces; 2) oleo resinous or lacquer, water repellent finish coat, sufficiently heavy and smooth to provide washable surface.

E. Concealed Portions: Finish with full flow coat of darkly pigmented, oleo resinous or lacquer, water repellent finish.

PART 3 - EXECUTION

3.01 INSTALLATION:

Install bookstack units at location shown, in continuous ranges made up of number of units shown, complying with manufacturer's instructions. Set units plumb and level, with permanent attachment and support.

Anchor single-faced ranges over 42" height to wall construction, by method recommended by this Standard.

3.02 Install shelves at spacings indicated or, if not indicated, at equal spacing in each unit.

3.03 Touch-up marred finishes or replace component parts as necessary to eliminate evidence of damage or deterioration.

METAL LIBRARY SHELVES

PART 1 - GENERAL

1.01 RELATED DOCUMENTS:

Drawings and general provisions of Contract, including General and Supplementary Conditions and Division-1 Specification sections, apply to work of this section.

1.02 RELATED WORK:

List those trades which supply backing and bracing materials for anchorage of shelving.

1.03 DESCRIPTION OF WORK:

Extent of library shelving is indicated on drawings.

Types of library shelving required include:

Adjustable metal bookshelf units.
Bracketed-type cantilevered metal bookstack units.

DELETE INAPPLICABLE TYPES FROM ABOVE LIST;
ADD OTHER TYPES AS REQUIRED.

1.04 QUALITY ASSURANCE:

Manufacturers: Provide each type of library shelving by one manufacturer for entire project.

Test of Enamel Finish: Provide testing of five random samples of enamel finish on sheet steel, taken from the work fabricated and finished for project. Arrange tests to be performed by an independent testing laboratory, complying with ASTM D 968, including analysis and report of results.

Regulatory Requirements: Provide bookstack design complying with structural and anchoring requirements for seismic Zone (3 or 4).

1.05 SUBMITTALS:

A. Product Data: Submit manufacturer's technical product literature and installation instructions for bookstack units.

B. Samples: Submit four metal panels 3" x 5" with specified finish and manufacturer's data and color palette for color selection.

C. Shop Drawings: Submit shop drawings for bookstacks, showing details, dimensions and layout of installation with verification of conformance with plans. Show details of anchorages and closure panels (if any).

D. Seismic Calculations: Submit detailed calculations of seismic forces for fully loaded book stacks, indicating compliance with seismic design requirements of Title 24, Part 2, CCR before fabrication.

1.06 ACCEPTABLE MANUFACTURERS:

Available Manufacturers: Subject to compliance with requirements, manufacturers offering products which may be incorporated in the work include, but are not limited to, the following:

RETAIN ABOVE OR BELOW, FOR NONPROPRIETARY OR SEMIPROPRIETARY SPECIFICATION.

Manufacturer: Subject to compliance with requirements, provide products of one of the following:

Aetnastak Div./Art Metal-U.S.A., Inc.
Gaylord Bros., Inc.
Library Bureau, Inc.
RCH-Spacemaster/Div. Reflector Hardware Corp.
Republic Steel Corp./Industrial Products Div.
Borroughs-Wilsonstak
All-Steel Canada, LTD
Richards-Wilcox

DELETE OR ADD MANUFACTURERS TO ABOVE LIST AS DESIRED

1.07 MATERIALS:

Sheet Steel: ASTM A 366, cold-rolled sheet, commercial quality, Class 1, matte finish.

Provide stretcher-leveled sheet for shelves and end panels.

Fasteners: Cadmium-plated steel, manufacturer's standard types and sizes.

REVISE BELOW IF "SPECIAL" FINISH REQUIRED.

Sheet Steel Finish: Manufacturer's standard baked enamel finish; minimum of 1.0 mils dry film thickness.

DELETE BELOW FOR MINOR WORK AND WHERE ENFORCEMENT IS UNLIKELY. (SEE RELATED "TESTING" PARAGRAPHS UNDER "SUBMITTALS" AND "QUALITY ASSURANCE" IN PART 1).

Provide coating with abrasion coefficient of 25 liters per mil when tested in accordance with ASTM D 968.

Provide color selected by Architect from manufacturer's standard colors.

PART 2 - FABRICATION

2.01 GENERAL:

Fabricate work in shop to greatest extent possible, before application of finish. Remove sharp and rough edges and corners from cut metal and grind welds smooth. Design components, joints and connections, to withstand most severe possible book loading condition, with normal factor of safety. Provide units of manufacturer's standard design and fabrication methods, but not lesser strength and quality than shown and specified.

Design units so that bookstacks in ranges indicated will be completely adjustable, without loss of stability when one or more shelves are removed at any location. Provide for adjustment of shelf heights at 1" increments, from top to bottom of units. Provide bookstacks of "unitized construction" such that an individual bookstack unit may be removed from a range and relocated without disturbing or altering stability of adjacent units.

CAREFULLY DELETE THE PROVISIONS FROM THE REMAINDER OF THIS ARTICLE, WHICH ARE NOT APPLICABLE TO THE TYPES OF UNITS REQUIRED.

2.02 UNIT SIZES:

A. Width: Except as otherwise indicated, provide units of standard 36" nominal width, with shelf depth and unit heights as follows:

B. Bracketed Shelf Depth: Nominal 8", 9", 10", or 12".

RETAIN ONLY ONE OF 4 ABOVE CHOICES.

C. Overall Height: 90".

REVISE ABOVE (STANDARD) IF NECESSARY (42", 66", 78", AND 102" ALSO AVAILABLE FROM MOST MFRS.).

2.03 STANDARDS:

Wall-Supported Posts: Manufacturer's standard, but not less than 16 gauge.

Upright Posts: Manufacturer's standard tubular of back-to-back channel design; size, gauge and reinforcement as needed for loading, but not less than 16 gauge. Provide base shoe with resilient pad of rubber or neoprene, except where base brackets or other components transmit loading to floor.

Bottom Spreaders: 18 gauge minimum, 16 gauge if welded.

Top Tie Channels: 18 gauge minimum, 16 gauge if welded.

Weld bottom spreaders and top tie channels to upright posts if necessary for elimination of longitudinal sway bracing on free-standing ranges (diagonal bracing not permitted).

ABOVE IS A SPECIAL FEATURE; AND IS ALSO NOT NEEDED FOR CEILING SUPPORTED OR WALL BRACED UNITS.

Extend tie channels at each range to supporting building construction, with anchorage as necessary for longitudinal support of range.

ABOVE IS ANOTHER (UNCOMMON) WAY OF ELIMINATING SWAY BRACING. A "RANGE" IS ROW OF UNITS.

Base Brackets: 16 gauge minimum for flanged construction; 13 gauge minimum for plate-type construction; 10" high x full width of base shelf; top and front edges rounded and flared to match shelf brackets. Provide adjustable leveling device at each end of bracket, with shoe and resilient pad of rubber or neoprene, located for concealment under base shelf and behind end panel.

BELOW IS ORDINARILY USED ONLY WITH BASE BRACKETS (FREE-STANDING RANGES).

Base Shelf and Base: 18 gauge minimum, manufacturer's standard closed-base construction.

BELOW NOT NORMALLY REQUIRED WITH WELDED FRAMES.

Sway Braces: Shall be bolted into uprights as required by engineer or certified tests.

Cross-Tie Channels: Minimum 18 gauge x 2" x 1" box-channel (4 bends), extended at each bay of each range to supporting building construction, with anchorage as necessary for lateral support of each unit in each range. Splice plates and connections shall be equal in strength to box channels.

ABOVE (AND BELOW) IS A METHOD OF ELIMINATING BASE BRACKETS FROM UNITS WHICH DO NOT EXTEND TO CEILING CONSTRUCTION.

At manufacturer's option, cross-tie channels may be spaced at every third bay (post row) if top tie channel (longitudinal) is continuous and reinforced.

Adjustable Bracketed Shelf: 20 gauge minimum shelf designed for 50 lbs. per sq. ft. loading with not more than 3/16" deflection; with 3-bend edge construction (both edges) designed to receive wire-type book supports. Minimum 16 gauge brackets with rounded and flared (1/4" thick) tops and ends to prevent book knifing, with 2 post-hooks near top and one safety-lug near bottom.

Shelf Quantities: Except where other shelf arrangements are shown or scheduled, provide the following number of adjustable bracketed shelves for each face of each unit.

EDIT THE FOLLOWING TO RETAIN ONLY APPLICABLE QUANTITIES.

Ceiling-High Units: 7 shelves (plus base).
90" High Units: 6 shelves (plus base).
66" High Units: 4 shelves (plus base).
42" High Units: 2 shelves (plus base).

2.04 ACCESSORIES AND SPECIAL SHELVES:

A. General: Provide manufacturer's standard accessory items and special-use shelf units of types indicated in quantities and locations indicated.

DELETE TYPES NOT NEEDED. INSERT OR REVISE SCOPE OF EACH ITEM.

B. Accessories: Provide V-type range finder on aisle-end of each range, for 5" x 3" cards.

Provide cast-aluminum card-holder on each face of end panels.

DELETE ABOVE IF STANDARD STEEL CARD SLOT IS ACCEPTABLE.

Provide 2 cast-aluminum card-holders on each end panel (single on wall units) which faces onto an aisle; for 5" x 3" cards.

Provide range finders of manufacturer's standard "V"-shaped design, designed to hold 3" x 5" cards, finished to match shelving. Provide one per exposed end panel.

Provide wire-type book supports of manufacturer's standard design, to clamp into the shelf above, 6 gauge wire with cadmium, chromium or nickel-plated finish.

Provide 2 supports for every 3 shelves.

Provide 5" long label holders of manufacturer's standard snap-on or slip-on design, for slip-in label strips.

Wood End Panels: Provide manufacturer's standard construction of hardwood veneer, of species and finish indicated, on plywood or particle-board core, with matching solid hardwood exposed edges.

DELETE ABOVE IF NONE; OTHERWISE, SHOW LOCATIONS, SPECIES, FINISH, ETC. ON DRAWINGS, OR INSERT FURTHER DESCRIPTION HERE. REVISE IF SOLID HARDWOOD PANELS DESIRED.

Plastic Laminate: NEMA LD-3, high pressure type, Grade GP-50 (0.050" thick), color and pattern as selected by Architect from manufacturer's standards, unless otherwise indicated.

DELETE ABOVE IF NOT DESIRED FOR END PANELS.

PART 3 - EXECUTION

3.01 INSTALLATION:

Install bookstack units at location shown, in continuous ranges made up of number of units shown, complying with manufacturer's instructions. Set units plumb and level, with permanent attachment and support.

Anchor single-faced ranges over 42" height to wall construction, by method recommended by this Standard.

Assemble support system (to whatever extent not factory assembled) by bolting and welding as required. Coordinate field welding (if any) with other work and safety requirements. Anchor supporting members (posts, tie channels, brackets) to walls, ceilings, columns or beams as required for stability of units.

Install end panels and canopy tops by concealed bolting.

Install shelves at spacings indicated or, if not indicated, at equal spacing in each unit.

Touch-up marred finishes or replace component parts as necessary to eliminate evidence of damage or deterioration.

MOVEABLE COMPACT LIBRARY SHELVING

PART 1 - GENERAL

1.01 RELATED DOCUMENTS:

Drawings and general provisions of Contract, including General and Supplementary Conditions and Division-1 Specification sections, apply to work of this section.

1.02 RELATED WORK:

List those trades which supply backing and bracing materials for anchorage of shelving.

1.03 DESCRIPTION OF WORK:

Extent of library shelving is indicated on drawings.

Types of library shelving required include:

- Adjustable metal bookshelf units.
- Bracketed-type cantilevered metal book stack units.
- High density mobile storage system.
  - Manual operation.
  - Mechanically assisted manual operation.
  - Electric operation (Microprocessor controlled).

DELETE INAPPLICABLE TYPES FROM ABOVE LIST; ADD OTHER TYPES AS REQUIRED.

1.04 QUALITY ASSURANCE:

Manufacturers: Provide each type of library shelving by one manufacturer for entire project. Shelving shall be furnished and installed only by firms engaged in the manufacture and installation of this type of equipment for the last five years.

Test of Enamel Finish: Provide testing of five random samples of enamel finish on sheet steel, taken from the work fabricated and finished for project. Arrange tests to be performed by an independent testing laboratory, complying with ASTM D 968, including analysis and report of results.

Regulatory Requirements: Provide book stack and carriage design complying with structural and anchoring requirements for Seismic Zone (3 or 4).

CHOOSE ONE

1.05 SUBMITTALS:

A. Product Data: Submit manufacturer's technical product literature and installation instructions for book stack units.

B. Samples: Submit four metal panels 3" x 5" with specified finish and manufacturer's data and color palette for color selection.

C. Shop Drawings: Submit shop drawings for book stacks and carriage complete, showing details, dimensions and layout of installation with verification of conformance with plans. Show details of anchorages and closure panels (if any).

D. Seismic Calculations: Submit detailed calculations of seismic forces for fully loaded book stacks, indicating compliance with seismic design requirements of Title 24, Part 2, CCR signed by a Civil or Structural Engineer licensed in California.

1.06 ACCEPTABLE MANUFACTURERS:

Available Manufacturers: Subject to compliance with requirements, manufacturers offering products which may be incorporated in the work include, but are not limited to, the following:

RETAIN ABOVE OR BELOW, FOR NONPROPRIETARY OR SEMIPROPRIETARY SPECIFICATION.

Manufacturer: Subject to compliance with requirements, provide products of one of the following:

Electropack Co. Ltd.
Tab Products Co.
RHC-Spacemaster/Div. Reflector Hardware Corp.
Space Saver

DELETE OR ADD MANUFACTURERS TO ABOVE LIST AS DESIRED

1.07 MATERIALS:

Sheet Steel: ASTM A 366, cold-rolled sheet, commercial quality, Class 1, matte finish.

Provide stretcher-leveled sheet for shelves and end panels.

Fasteners: Cadmium-plated steel, manufacturer's standard types and sizes.

REVISE BELOW IF "SPECIAL" FINISH REQUIRED.

Sheet Steel Finish: Manufacturer's standard baked enamel finish; minimum of 1.0 mils dry film thickness.

DELETE BELOW FOR MINOR WORK AND WHERE ENFORCEMENT IS UNLIKELY. (SEE RELATED "TESTING" PARAGRAPHS UNDER "SUBMITTALS" AND "QUALITY ASSURANCE" IN PART 1).

Provide coating with abrasion coefficient of 25 liters per mil when tested in accordance with ASTM D 968.

Provide color selected by Architect from manufacturer's standard colors.

PART 2 - FABRICATION

2.01 GENERAL:

Fabricate work in shop to greatest extent possible, before application of finish. Remove sharp and rough edges and corners from cut metal and grind welds smooth. Design components, joints and connections, to withstand most severe possible book loading condition, with normal factor of safety. Provide units of manufacturer's standard design and fabrication methods, but not lesser strength and quality than shown and specified.

Design units so that book stacks in ranges indicated will be completely adjustable, without loss of stability when one or more shelves are removed at any location. Provide for adjustment of shelf heights at 1-1/2" increments, from top to bottom of units. Provide book stacks of "unitized construction" such that an individual book stack unit may be removed from a range and relocated without disturbing or altering stability of adjacent units.

CAREFULLY DELETE THE PROVISIONS FROM THE REMAINDER OF THIS ARTICLE, WHICH ARE NOT APPLICABLE TO THE TYPES OF UNITS REQUIRED.

2.02 UNIT SIZES:

A. Width: Except as otherwise indicated, provide units of standard 36" nominal width, with shelf depth and unit heights as follows:

B. Bracketed Shelf Depth: Nominal 8", 9", 10", or 12".

RETAIN ONLY ONE OF 4 ABOVE CHOICES.

C. Overall Height: 90".

REVISE ABOVE (STANDARD) IF NECESSARY (42", 66", 78", AND 102" ALSO AVAILABLE FROM MOST MFRS.).

2.03 STANDARDS:

Wall-Supported Posts: Manufacturer's standard, but not less than 16 gauge.

Upright Posts: Manufacturer's standard tubular of back-to-back channel design; size, gauge and reinforcement as needed for loading, but not less than 16 gauge.

Base Brackets: 16 gauge minimum for flanged construction; 13 gauge minimum for plate-type construction; 10" high x full width of base shelf; top and front edges rounded and flared to match shelf brackets.

Base Shelf and Base: 18 gauge minimum, manufacturer's standard closed-base construction.

BELOW NOT NORMALLY REQUIRED WITH WELDED FRAMES.

Sway Braces: As required by engineer or certified tests. Shall be bolted to uprights.

Adjustable Bracketed Shelf: 20 gauge minimum shelf designed for 50 lbs. per sq. ft. loading with not more than 3/16" deflection; with 3-bend edge construction (both edges) designed to receive wire-type book supports. Minimum 16 gauge brackets with rounded and flared (1/4" thick) tops and ends to prevent book knifing, with 2 post-hooks near top and one safety-lug near bottom.

Shelf Quantities: Except where other shelf arrangements are shown or scheduled, provide the following number of adjustable bracketed shelves for each face of each unit.

EDIT THE FOLLOWING TO RETAIN ONLY APPLICABLE QUANTITIES.

Ceiling-High Units: 7 shelves (plus base).
90" High Units: 6 shelves (plus base).
66" High Units: 4 shelves (plus base).
42" High Units: 2 shelves (plus base).

2.04 ACCESSORIES AND SPECIAL SHELVES:

A. General: Provide manufacturer's standard accessory items and special-use shelf units of types indicated in quantities and locations indicated.

DELETE TYPES NOT NEEDED. INSERT OR REVISE SCOPE OF EACH ITEM.

B. Accessories: Provide V-type range finder on aisle-end of each range, for 5" x 3" cards.

Provide cast-aluminum card-holder on each face of end panel.

DELETE ABOVE IF STANDARD STEEL CARD SLOT IS ACCEPTABLE.

Provide 2 cast-aluminum card-holders on each end panel (single on wall units) which faces onto an aisle; for 5" x 3" cards.

Provide range finders of manufacturer's standard "V"-shaped design, designed to hold 3" x 5" cards, finished to match shelving. Provide one per exposed end panel.

Provide wire-type book supports of manufacturer's standard design, to clamp into the shelf above, 6 gauge wire with cadmium, chromium or nickel-plated finish.

Provide 2 supports for every 3 shelves.

Provide 5" long label holders of manufacturer's standard snap-on or slip-on design, for slip-in label strips.

C. Wood End Panels: Provide manufacturer's standard construction of hardwood veneer, of species and finish indicated, on plywood or particle-board core, with matching solid hardwood exposed edges.

DELETE ABOVE IF NONE; OTHERWISE, SHOW LOCATIONS, SPECIES, FINISH, ETC. ON DRAWINGS, OR INSERT FURTHER DESCRIPTION HERE. REVISE IF SOLID HARDWOOD PANELS DESIRED.

Plastic Laminate: NEMA LD-3, high pressure type, Grade GP-50 (0.050" thick), color and pattern as selected by Architect from manufacturer's standards, unless otherwise indicated.

DELETE ABOVE IF NOT DESIRED FOR END PANELS.

2.05 CARRIAGE:

A. Track:

The track assembly shall be designed and installed to adequately distribute the carriage loads to the existing floor and to provide guidance for the carriage wheels. The rails shall be steel as specified by the engineer, with a top surface of not less than 5/8". The top surface shall be machined for smooth operation, and all track sections tongue and groove. The track for the entire system shall be completely leveled along the length of track and with each other to within 1/32" and anchored to the floor. If welding is required all welds shall be performed by certified welders and inspected by a certified testing laboratory.

B. Raised Deck:

The entire system shall be equipped with a low profile floor beneath the entire range lengths and between the track assembly system so that it will be flush with the top surface of the tracks. The deck shall be a minimum 5/8" thick AD plywood and shall be adequately braced to prevent deflection and lateral movement.

Floor shall have integrated leveling screws minimum 16" on center. A ramp of the same material as floor at the front of each assembly should not exceed a slope of 1" in 6" at the main access aisle (which resembles the letter "T" due to the arrangement of the shelving units) and shall not extend past the end of the units into the main access aisle. The plywood floor and ramps shall be covered with commercial grade tile 1/8" such as Azrock vinyl, Armstrong Solarian, or approved equal.

C. Wheel Assemblies and Carriages:

The steel wheels, carriages, and drive mechanisms shall accommodate a total load of not less than 1,000 pounds per carriage foot in each range and shall accommodate unusual loads with a proper safety factor.

Wheel Assemblies: Wheels shall be a minimum of 2-1/2" diameter. Wheels to be machined high grade steel, hardened, with 3/4" minimum diameter solid steel axles and two 2,000 pound permanently lubricated bearing assemblies per wheel. All bearings shall be permanently sealed and self aligning. Drive shafts shall be 1" diameter solid steel running continuously the full length of each range. All splices and connections between drive shafts and axles shall be by means of securely retained rigid key way connections. Complete drive shaft and wheel assemblies shall exhibit no play or looseness over the entire length of that assembly.

Carriage Assemblies: The carriages are to have main supporting structural sections with 5/8" high shelf retaining flanges running the full length of the carriage. A second section is positioned and welded between the main support sections. The carriages are to be a one-piece welded member capable of supporting a minimum load of 1,000 lbs. per carriage length foot. They are to be designed to prevent tipping-over of units.

D. Aisle Lock:

Each movable range shall be equipped with an automatic aisle locking mechanism or alternative system to prevent harm to people.

PART 3 - EXECUTION

3.01 INSTALLATION:

Install system at the locations and groupings shown, and shall be set plumb, level and square. All components and sections, to whatever extent not factory assembled, shall be site assembled and installed by clipping, bolting and welding as required. Field welding shall meet safety regulations and be comparable to shop quality and standards.

Installation of the rails, carriages and platforms shall be under the direct supervision of factory trained personnel.

Install shelves at spacings indicated or, if not indicated at equal spacing in each unit. All marred finishes of materials/equipment supplied by contractor shall be touched up or otherwise repaired to eliminate any source of injury to people or books in normal use.

3.02 DELIVERY:

All items shall be suitably protected during shipment to prevent loss or damage. Truck parking space will be made available by the Library only for the nominal period required to complete unloading (advance notice required).

The contractor shall provide all equipment and tools necessary to unload the trucks, unwrap or uncarton the items, transfer them into site, and assembly and install as required. Care must be observed so as not to damage the walls, etc. during assembly and installation.

3.03 WORKMANSHIP:

Workmanship in the manufacture and installation of the system shall be of the high quality found in large research libraries where hard daily usage over several decades is typical without significant deterioration due to use. Joints and connections shall be well formed and tight. All sharp projections, edges and corners shall be removed so that persons and books will not be injured by contact with the system in normal use.

3.04 CLEAN-UP:

All dunnage shall be removed from library property. The shelving shall be wiped clean and floor left in broom clean condition so that staff can use and the books can occupy the space without further cleaning.

3.05 WARRANTY:

The vendor shall guarantee that the system and its installations shall be as specified and shall further guarantee them against defects in materials and workmanship for a period of not less than three (3) years following owner acceptance. During the three year period, the vendor shall repair or replace defective items and installation deficiencies, within 30 days of written notice by the owner, without further cost to the owner. In the event the vendor fails to make the corrections, the owner may proceed to have the work done at the vendor's expense and the vendor shall honor and agree to pay all costs and charges. Normal wear and tear, vandalism alteration or damage shall not apply.

3.06 INSTRUCTIONS:

Provide operating and maintenance instructions. Also provide one training session for owner's operating and maintenance personnel if requested by owner within the warranty period.

3.07 ACCEPTANCE:

A. Independent Expert:

The owner shall reserve the option of contracting with an independent expert to examine the compact shelving installation to ensure it has been constructed to specification within thirty (30) days of completion of project.

B. Test Run:

Several ranges of compact shelving shall be loaded with the maximum number of books/materials in order that the system demonstrate its capabilities and to determine if it operates as specified.

APPENDIX "D"

MITIGATION OF EARTHQUAKE HAZARDS IN EXISTING LIBRARIES

MITIGATION OF EARTHQUAKE HAZARDS IN EXISTING LIBRARIES

The following materials are a brief selection of items in the extensive reference collection of the California Seismic Safety Commission, Sacramento. They are selected primarily for their recency, and because they address typical concerns in existing libraries. Most are available from libraries, or from on line, or entrepreneurial document delivery services. They can be a useful starting point to complement mitigation of book stack installations being performed with the help of "Seismic Safety Standards for Library Shelving, Manual of Recommended Practice" Sacramento, California State Library, 1989. All information below is take from the publications themselves.

-Cy H. Silver, Facilities Planning Consultant

1. Earthquake Engineering Research Foundation, "Non-structural issues on seismic design and construction; a technical evaluation based on a workshop sponsored by the National Science Foundation" Berkeley, CA, EERI, June 1984. (EERI Publication No. 84-04; NSF Grant No. CEE-8203510.) EERI's address is 2620 Telegraph Avenue, Berkeley, 94704.

2. "Earthquakes and Elevators", 1972 "Elevator World" October annual issue. The journal's address is P.O. Box 6523, Mobile, Alabama.

3. FIMS, Inc., "Data Processing Facilities, Guidelines for Earthquake Hazard Mitigation", prepared by FIMS, Inc. in cooperation with VSP Associates, Inc., June 1987. Available from VSP Associates, Inc., 2222 Watt Avenue, Suite D-4, Sacramento, CA, phone (916) 971-9220.

4. Kockelman, William J., "Reducing Losses from Earthquakes Through Personal Preparedness". U.S. Geological Survey, 1984. (Open-file Report 84-765.)

5. Lagorio, Henry J., "Issues for Seismic Strengthening of Existing Buildings: A Practical Guide for Architects", [by] Henry J. Lagorio, Howard Friedman, [and] Kit Wong. Berkeley, CA, University of California Center for Environmental Design Research, January 1986. "Dr. S.C. Liu, Project Director, National Science Foundation."

6. "Reducing the Risks of Non-structural Earthquake Damage: A Practical Guide", 2d ed. Oakland, CA, Bay Area Regional Earthquake Preparedness Project, October 1985. (BAREPP 85-10) "Originally developed [and published in 1983 -comp.] by Robert Reitherman of Scientific Service, Inc., for the Southern California Earthquake Preparedness Project." "The revisions [to the Reitherman work]...consist primarily of modifying graphics, updating construction cost estimates, and identifying the need for engineering and architectural assistance..."BAREPP's address is 101 - 85h Street, Suite 152, Oakland, 90607; phone (415) 540-2713. [Compiler's note: this work appears to be the same as the of the title by FEMA.]

6a. "Checklist of Non-structural Earthquake Hazards", Bay Area Regional Earthquake Preparedness Project [n.d.] "This list should be used in conjunction with "Reducing the Risks of Non-structural Earthquake Damage: A Practical Guide", BAREPP's "how to guide to reducing non-structural hazards in the home and work place."

7. "Reducing the Risks of Non-structural Earthquake Damage: A Practical Guide". U.S. Federal Emergency Management Agency, June 1985. "Earthquake Hazards Reduction Series 1." (FEMA 74/June 1985.) With inserted "update 12/85". [Compiler's note: this work, with its enclosed update, appears to be the same as that of the same title by BAREPP.]

8. Repair and retrofit of existing structures, 1987, Los Angeles. "Repair and Retrofit of Existing Structures". Austin, TX, Phil M. Ferquson Structural Engineering Laboratory, Department of Civil Engineering, Bureau of Engineering Research, University of Texas, December, 1987. "Report of workshop, Los Angeles, CA, September 30 - October 1, 1987. Sponsored by National Science Foundation Grant No. ECE-8701316." (PMFSEL Report No. 87-6).

9. "Seismic Design Guidelines for Upgrading Existing Buildings; Technical Manual". U.S. Departments of the Army, Navy, and the Air Force, September 1, 1988. (Army TM 5-809-10-2; Navy NAVFACP-355.2; Air Force AFM 88-3, Chapter 13, Section B.)

10. University of California, College of Environmental Design, Center for Planning and Development Research, "An Earthquake Advisor's Handbook for Wood Frame Houses", Berkeley, Summer 1982.